BE AGILE®

Agile for Executives

Part of the Agile Education Series™

CAPE PROJECT
MANAGEMENT, INC.

Copyright Notice 2019.

▸ Unless otherwise noted, these materials and the presentation of them are based on the methodology developed by Cape Project Management, Inc. and are copyrighted materials of Cape Project Management, Inc., which are used with its permission. Cape Project Management, Inc. reserves all its rights.

▸ All other brand or product names used in this guide are trade names or registered trademarks of their respective owners.

▸ Portions of this slide deck are attributable to Mountain Goat Software, LLC under Creative Commons.

Version 3.1

CAPE PROJECT
MANAGEMENT, INC.

2

About Us

- ▸ The course curriculum developed by Dan Tousignant, of Cape Project Management, Inc.
- ▸ We provide public, onsite and online training:
  - ◦ Agile.Us.com
- ▸ Follow us on Twitter @ScrumDan
- ▸ The content of this course is licensed to your instructor.

CAPE PROJECT
MANAGEMENT, INC.

3

The Agile Education Series™

1. Scrum Master Certification Training
2. Product Owner and User Story Training
3. All About Agile™: PMI-ACP® Agile Exam Preparation
4. Kanban for Software Development Teams
5. Achieving Agility – How to implement Agile in your organization
6. Agile for Team Members
7. Agile for Executives
8. DevOps for Agile Teams

▸ All of these curriculums are available on Amazon at:
  ◦ http://bit.ly/DansAgileBooks

4   Copyrighted materials. 2019   CAPE PROJECT MANAGEMENT, INC.

Agile Ice Breaker

The
Marshmallow
Challenge

20 sticks of spaghetti   +   one yard tape   +   one yard string + one marshmallow

5

CAPE PROJECT
MANAGEMENT, INC.

Notes:

Introductions and Expectations

6

CAPE PROJECT
MANAGEMENT, INC.

Notes:

Course Objectives

- ▸ Understand the key principles of Agile
- ▸ Learn the best practices of Scrum
- ▸ Recognize the management challenges and opportunities of implementing Agile

CAPE PROJECT
MANAGEMENT, INC.

7

Notes:

Agenda

Modules
1. Introduction to Agile
2. The Scrum Framework
3. The Agile Product Lifecycle
4. Agile Controls
5. Implementing Agile

8

Copyrighted materials. 2019

CAPE PROJECT
MANAGEMENT, INC.

Notes:

Announcements

- ‣ Participant materials
  - ∘ Slides
  - ∘ Exercises
- ‣ Breaks

9   Copyrighted materials. 2019   CAPE PROJECT MANAGEMENT, INC.

Notes:

Agile

Lean   Kanban
      XP   RUP
Crystal   Scrum

Introduction to Agile
Module 1

10

CAPE PROJECT
MANAGEMENT, INC.

Notes:

DISCUSSION

Why do projects fail?

CAPE PROJECT
MANAGEMENT, INC.

11

Notes:

Typical Project Risks

▸ Misunderstanding of the requirements

▸ Lack of management commitment and support

▸ Lack of adequate user involvement

▸ Failure to gain user commitment

▸ Failure to manage end user expectation

▸ Changes to requirements

▸ Lack of an effective project management methodology

http://mooc.ee/MTAT.03.243/2015_spring/uploads/Main/top-10.pdf

12   Copyrighted materials. 2019   CAPE PROJECT MANAGEMENT, INC.

Notes:

Project Success Rates

Waterfall	11%	60%	29%
Agile	39%	52%	9%

Successful   Challenged   ■ Failed

Success= On-time, On-budget with a satisfactory result
(% of requirements met strategic goal)

https://www.infoq.com/articles/standish-chaos-2015

CAPE PROJECT
MANAGEMENT, INC.

13

Notes:

Why is Agile more successful?

- Allows for more flexibility in requirements and development
- Includes built-in increments and iterations
- Provides the opportunity to encounter and address errors sooner in the development cycle
- Increases organizational and team efficiency
- Decreases unnecessary documentation and meetings
- Provides a value-based approach to development
- Assumes organizational differences
  - "Can be right sized"

14

CAPE PROJECT
MANAGEMENT, INC.

Notes:

Reasons for Adopting Agile

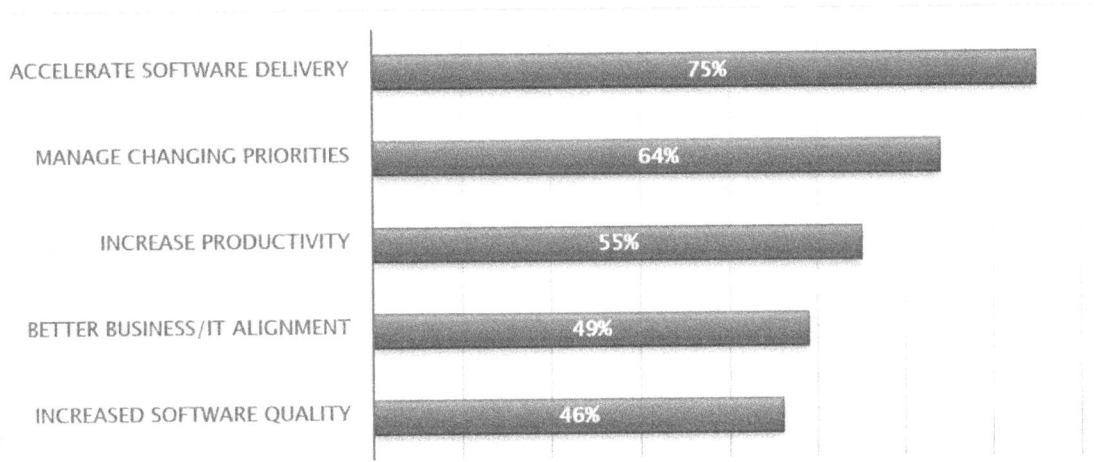

Reason	Percentage
ACCELERATE SOFTWARE DELIVERY	75%
MANAGE CHANGING PRIORITIES	64%
INCREASE PRODUCTIVITY	55%
BETTER BUSINESS/IT ALIGNMENT	49%
INCREASED SOFTWARE QUALITY	46%

Source: Version One 12th Annual Agile Survey, 2018

CAPE PROJECT
MANAGEMENT, INC.

15

Notes:

When to use Agile:

Requirements Stability vs. Development Approach

- Selecting the right project approach depends upon the stability of the requirements:
  - A predictive team can report exactly what features and tasks are planned for the entire length of the development process and can commit to a fixed cost, schedule and scope.
  - Adaptive methods focus on adapting quickly to changing realities. When the needs of a project change, an adaptive team changes as well.

Predictive   **Adaptive**

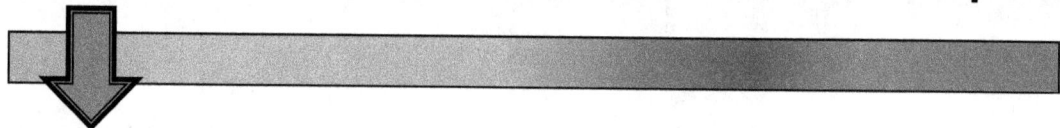

Traditional/   Incremental   Iterative   Agile
Waterfall

16   Copyrighted materials. 2019   CAPE PROJECT
                                                                     MANAGEMENT, INC.

Notes:

DISCUSSION

Why be more Agile or Adaptive in your organization?

What types of requirement changes do you experience?

CAPE PROJECT
MANAGEMENT, INC.

17

Notes:

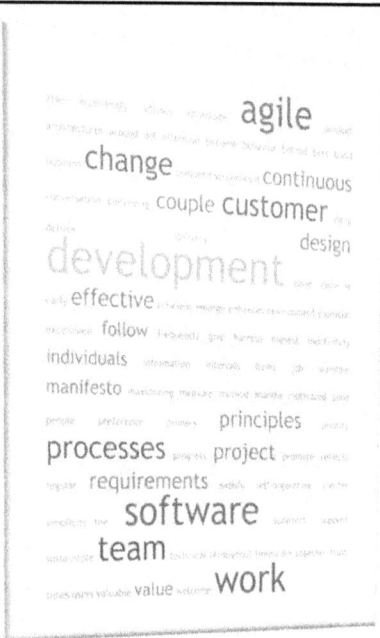

The
Agile Manifesto

18

CAPE PROJECT
MANAGEMENT, INC.

What does it mean to "Be Agile?"

- The term Agile comes from the Agile Manifesto
  - 4 values
  - 12 principles
- In summary: 4 takeaways
  - Fast – as fast as is appropriate
  - Small empowered teams
  - Value driven delivery
  - Continuous Improvement

CAPE PROJECT
MANAGEMENT, INC.

19

Notes:

The Agile Manifesto: A statement of values

We are uncovering better ways of developing software by doing it and helping others do it.

Through this work we have come to value:

Individuals and interactions	over	Process and tools
Working software	over	Comprehensive documentation
Customer collaboration	over	Contract negotiation
Responding to change	over	Following a plan

That is, while there is value in the items on the right, we value the items on the left more.

Source: www.agilemanifesto.org

20

CAPE PROJECT MANAGEMENT, INC.

Notes:

12 Principles of the Agile Manifesto

1. Our highest priority is to satisfy the customer through early and continuous delivery of valuable software.

2. Welcome changing requirements, even late in development. Agile processes harness change for the customer's competitive advantage.

3. Deliver working software frequently, from a couple of weeks to a couple of months, with a preference to the shorter timescale.

4. Business people and developers must work together daily throughout the project.

5. Build projects around motivated individuals. Give them the environment and support they need, and trust them to get the job done.

6. The most efficient and effective method of conveying information to and within a development team is face-to-face conversation.

7. Working software is the primary measure of progress.

8. Agile processes promote sustainable development. The sponsors, developers, and users should be able to maintain a constant pace indefinitely.

9. Continuous attention to technical excellence and good design enhances agility.

10. Simplicity—the art of maximizing the amount of work not done—is essential.

11. The best architectures, requirements, and designs emerge from self-organizing teams.

12. At regular intervals, the team reflects on how to become more effective, then tunes and adjusts its behavior accordingly.

CAPE PROJECT
MANAGEMENT, INC.

Notes:

Activity

Your 3 Principles

CAPE PROJECT
MANAGEMENT, INC.

Notes:

Activity: Agile Manifesto Principles

Directions:

1. Pair-up with another person at your table and review the Agile Manifesto Principles.
2. Pick three principles that you think are critical to the success of your Agile Implementation or are the most challenging.
3. Be prepared to share your answer with the class.

Principle	Choose 3
1. Our highest priority is to satisfy the customer through early and continuous delivery of valuable software.	
2. Welcome changing requirements, even late in development. Agile processes harness change for the customer's competitive advantage.	
3. Deliver working software frequently, from a couple of weeks to a couple of months, with a preference to the shorter timescale.	
4. Business people and developers must work together daily throughout the project.	
5. Build projects around motivated individuals. Give them the environment and support they need, and trust them to get the job done.	
6. The most efficient and effective method of conveying information to and within a development team is face-to-face conversation.	
7. Working software is the primary measure of progress.	

8. Agile processes promote sustainable development. The sponsors, developers, and users should be able to maintain a constant pace indefinitely.	
9. Continuous attention to technical excellence and good design enhances agility.	
10. Simplicity--the art of maximizing the amount of work not done--is essential.	
11. The best architectures, requirements, and designs emerge from self-organizing teams.	
12. At regular intervals, the team reflects on how to become more effective, then tunes and adjusts its behavior accordingly.	

Notes:

_____

_____

_____

_____

_____

_____

_____

Agile Framework Adoption

Framework	Percentage
Scrum	56%
Custom Hybrid	14%
Scrumban	8%
Scrum/XP Hybrid	6%
Other	6%
Kanban	5%
Iterative	3%
Spotify Model	1%
Lean Startup	1%
XP	1%

Source: Version One 12th Annual Agile Survey, 2018

Copyrighted materials. 2019

CAPE PROJECT MANAGEMENT, INC.

25

Notes:

Questions

CAPE PROJECT
MANAGEMENT, INC.

Notes:

The Scrum Framework
Module 2

CAPE PROJECT
MANAGEMENT, INC.

Scrum

▸ Scrum (n): A framework within which people can address complex adaptive problems, while productively and creatively delivering products of the highest possible value.

Source: Schwaber, Sutherland: *A Scrum Guide*

28

CAPE PROJECT
MANAGEMENT, INC.

Notes:

Scrum is:

- ▸ Scrum is a lightweight, simple to understand (but difficult to master) agile process framework.
- ▸ Scrum is one of several agile software development methods.
- ▸ Scrum and Extreme Programming (XP) are probably the two best-known Agile methods. XP emphasizes technical practices such as pair programming and continuous integration. Scrum emphasizes management practice such as the role of Scrum Master.
- ▸ Many companies use the management practices of Scrum with the technical practices of XP.

29   Copyrighted materials. 2019   CAPE PROJECT MANAGEMENT, INC.

Notes:

Scrum is Agile

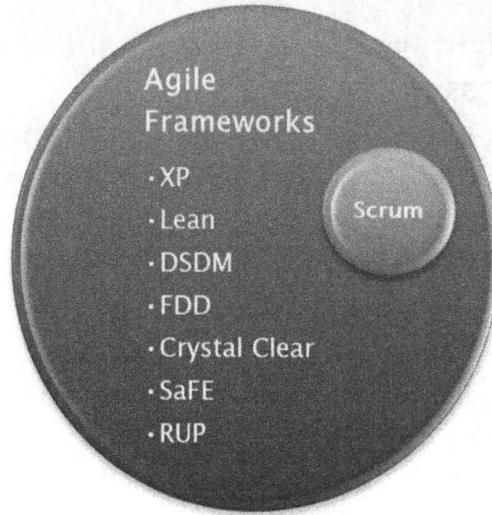

Agile Frameworks

- XP
- Lean
- DSDM
- FDD
- Crystal Clear
- SaFE
- RUP

Scrum

30

CAPE PROJECT
MANAGEMENT, INC.

Notes:

3 Pillars of Scrum

Transparency

▸ All relative aspects of the process must be visible to those responsible for the outcome.

Inspection

▸ There is frequent inspection of the artifacts and progress to identify and correct undesirable variances. Inspection occurs during the Sprint Planning Meeting, Daily Scrum, Sprint Review and Sprint Retrospective.

Adaptation

▸ After inspection, adjustments should be made to the processes and artifacts to minimize further deviation.

CAPE PROJECT
MANAGEMENT INC.

31

Notes:

5 Core Values of Scrum

1. **Commitment** – When we, as a team, value the commitment we make to ourselves and our teammates, we are much more likely to give our all to meet our goals.

2. **Focus** – When we value Focus, and devote the whole of our attention to only a few things at once, we deliver a better quality product, faster.

3. **Openness** – When we value being Open with ourselves and our teammates, we feel comfortable inspecting our behavior and practices, we can adapt them accordingly.

4. **Respect** – When we value Respect, people feel safe to voice concerns and discuss issues, knowing that their voices are heard and valued.

5. **Courage** – When we value Courage, people are encouraged to step outside of their comfort zones and take on greater challenges, knowing they will not be punished if they fail.

CAPE PROJECT
MANAGEMENT, INC.

Notes:

Scrum Project Management

A Product Owner creates a prioritized wish list (product backlog).

During Sprint Planning, the team pulls a small chunk from the top of that wish list (Sprint Backlog) and decides how to develop those pieces.

At the end of the Sprint, the work should be shippable (ready to hand to a customer, put on a store shelf, or show to a stakeholder).

Product Backlog

Sprint Backlog

Sprint

Working Software

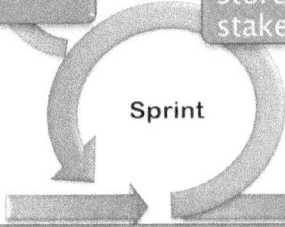

The team has a certain amount of time (Sprint) to complete its work, but meets each day to assess its progress, a Scrum.

Copyrighted materials. 2019

CAPE PROJECT MANAGEMENT, INC.

33

Notes:

The Scrum Framework

Roles
- Product Owner
- Scrum Master
- Team

Events
- Sprint Planning
- Sprint Review
- Sprint Retrospective
- Daily Scrum

Artifacts
- Product backlog
- Sprint backlog
- Burndown charts

34

CAPE PROJECT
MANAGEMENT, INC.

Scrum Roles

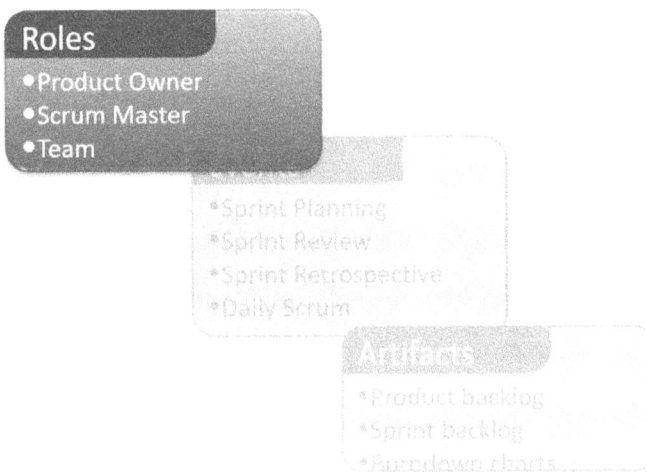

Roles
- Product Owner
- Scrum Master
- Team

- Sprint Planning
- Sprint Review
- Sprint Retrospective
- Daily Scrum

Artifacts
- Product backlog
- Sprint backlog
- Burndown charts

35

CAPE PROJECT
MANAGEMENT, INC.

The Scrum Master Dos

- ‣ Represents management to the project
- ‣ Responsible for enacting Scrum values and practices
- ‣ Removes impediments
- ‣ Ensure that the team is fully functional and productive
- ‣ Enable close cooperation across all roles and functions
- ‣ Shield the team from external interferences
- ‣ Is a Servant Leader

36

CAPE PROJECT
MANAGEMENT, INC.

Notes:

Servant Leadership

- ▸ Traditional leadership is "command-and-control"
  - ◦ "Workers need to be monitored closely"
- ▸ Servant leadership is based upon trust
  - ◦ "Team members are self-motivated"
- ▸ An agile servant leader needs to:
  - ◦ Protect the team from outside distractions
  - ◦ Remove impediments to the team's performance
  - ◦ Facilitates the team to address the tasks and resolve problems
  - ◦ "Move boulders and carry water"—in other words, remove obstacles that prevent the team from providing business value

CAPE PROJECT
MANAGEMENT, INC.

Notes:

Scrum Master Don'ts

- ‣ Own the product decisions on Product Owner's behalf
- ‣ Make estimates on team's behalf
- ‣ Make the technology decisions on team's behalf
- ‣ Assign the tasks to the team members
- ‣ Try to manage the team

38

CAPE PROJECT
MANAGEMENT, INC.

Notes:

The Product Owner Is a:

▸ Subject Matter Expert
- Understand the domain well enough to envision a product
- Answer questions on the domain for those creating the product

▸ End User and Customer Advocate
- Describe the product with understanding of users and its use.
- Understand the needs of the business and select a mix of features valuable to the customer

▸ Business Advocate
- Understand the needs of the organization paying for the software's construction and select a mix of features that serve their goals

▸ Communicator
- Capable of communicating vision and intent – deferring detailed feature and design decisions to be made just in time

▸ Decision Maker
- Given a variety of conflicting goals and opinions, be the final decision maker for hard product decisions

The Product Owner role is generally filled by a single person supported by a collaborative team

CAPE PROJECT MANAGEMENT, INC.

Notes:

Product Owner Dos

- ▸ Define the features of the product
- ▸ Decide on release dates and content
- ▸ Be responsible the total cost of ownership (TCO) and the profitability of the product (ROI)
- ▸ Orders features according to market value
- ▸ Adjust features and priority every Sprint, as needed
- ▸ Available daily* to answer questions for the Development Team

*not necessarily full-time, but as needed

CAPE PROJECT
MANAGEMENT, INC.

Notes:

Product Owner Don'ts

- Choose how much work will be accomplished in the Sprint - the team will do this, based on the priorities
- Change anything within the Sprint once it has started and don't add items unless the Sprint will end early
- Answer the three questions at the Daily Scrum Meetings (unless they have a task on a User Story)

CAPE PROJECT
MANAGEMENT, INC.

41

Notes:

Product Owner vs Scrum Master

Two different roles that complement each other. If one is not played properly, the other suffers.

▸ The Product Owner is responsible for the **product success**

▸ The Scrum Master is responsible for **project success**

42

Copyrighted materials. 2019

CAPE PROJECT
MANAGEMENT, INC.

Notes:

The Team Dos

- ▸ Typically 6 ± 3 people
- ▸ Teams are self-organizing
- ▸ Cross-functional:
  - ○ Programmers, testers, user experience designers, etc.
- ▸ Members should be full-time
  - • May be exceptions (e.g., database administrator)
- ▸ Team composition does not change during Sprint

CAPE PROJECT
MANAGEMENT, INC.

43

Notes:

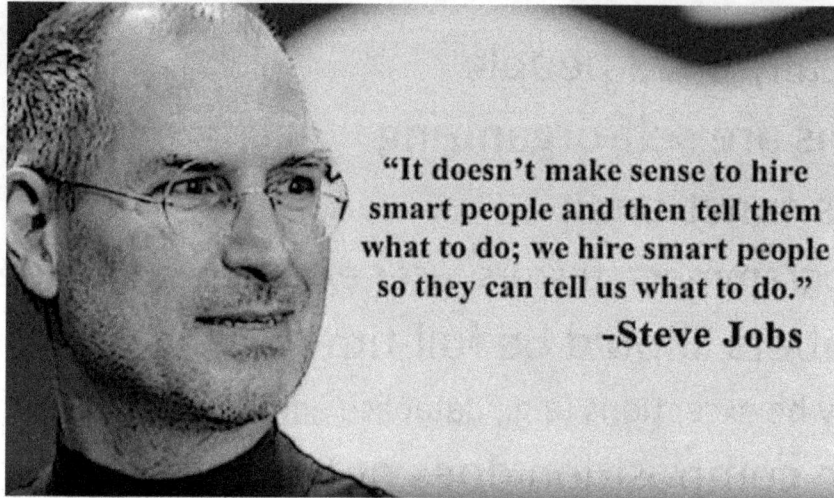

"It doesn't make sense to hire smart people and then tell them what to do; we hire smart people so they can tell us what to do."

-Steve Jobs

44

CAPE PROJECT
MANAGEMENT, INC.

Notes:

| |
| |
| |
| |
| |
| |
| |
| |

What is a self-organizing team?

- ▸ Team members are self-empowered (they know what needs to be done and have the means to do it)
- ▸ They are willing to take on the responsibility of self-organizing and self-examining
- ▸ They are ready to let go of their individual egos and differences so that the team can function together
- ▸ Scrum's success is dependent on the team's ability to self-examine itself and to continually seek improvement

http://blog.openviewpartners.com/scrum-challenge-self-organizing-teams/

45

CAPE PROJECT
MANAGEMENT, INC.

Notes:

The Team Don'ts

- Skip or cancel Scrum events
- Stop working when there is a roadblock or not enough information
- Increase technical debt in order to meet the velocity
- Individuals should not do excessive individual overtime, or in any other way try to be the "hero" of the team.

46

CAPE PROJECT
MANAGEMENT, INC.

Notes:

GROUP EXERCISE

The role of management in Agile

Start   Stop   Continue

CAPE PROJECT
MANAGEMENT, INC.

Notes:

Scaling Scrum

- Productivity comes from small teams
- If you have more than 9 people developing a single product, you need multiple teams
- The management of multiple Scrum teams is called "scaling Scrum"
- There are many different techniques for managing scaled Scrum. The most common elements among them are:
  - Dependency management via a Scrum of Scrums
  - A single Product Backlog and (chief) Product Owner

48

Copyrighted materials. 2019

CAPE PROJECT
MANAGEMENT, INC.

Notes:

The Scrum of Scrums

Scrum of Scrums Master

49

CAPE PROJECT
MANAGEMENT, INC.

Notes:

Scrum Events

Roles
- Product Owner
- Scrum Master
- Team

Events
- Sprint Planning
- Sprint Review
- Sprint Retrospective
- Daily Scrum

Artifacts
- Product backlog
- Sprint backlog
- Burndown charts

CAPE PROJECT MANAGEMENT, INC.

Scrum Events

Backlog Refinement can take up to 10% of the each Sprint. This involves creating new requirements and prioritizing and estimating them.

The **Daily Scrum** is 15 Minutes every day. Same time, same place.

Sprints are a maximum length of one-month. All work needs to meet a definition of "Done".

Scrum

During each **Sprint Review,** "Done" work is shown to stakeholders.

Product Backlog

Sprint Backlog

Sprint

Working Software

Sprint Planning occurs every Sprint. Half the session is to review requirements and half is for design.

Sprint Retrospectives ensure the Team is focusing on continuous improvement.

51

Copyrighted materials. 2019

MANAGEMENT, INC.

Notes:

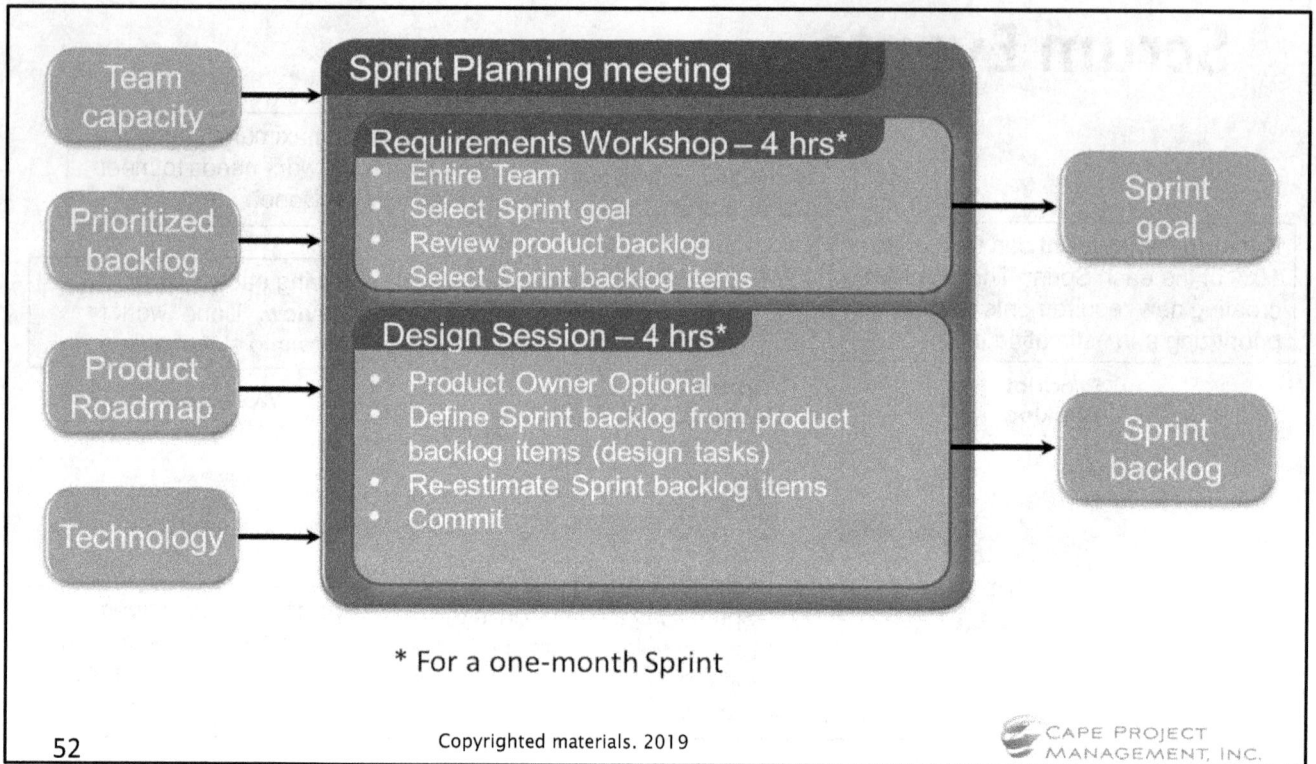

Sprint Planning meeting

Requirements Workshop – 4 hrs*
- Entire Team
- Select Sprint goal
- Review product backlog
- Select Sprint backlog items

Design Session – 4 hrs*
- Product Owner Optional
- Define Sprint backlog from product backlog items (design tasks)
- Re-estimate Sprint backlog items
- Commit

Inputs: Team capacity, Prioritized backlog, Product Roadmap, Technology

Outputs: Sprint goal, Sprint backlog

* For a one-month Sprint

CAPE PROJECT MANAGEMENT, INC.

52

Notes:

Questions

53

CAPE PROJECT
MANAGEMENT, INC.

Notes:

Scrum Artifacts

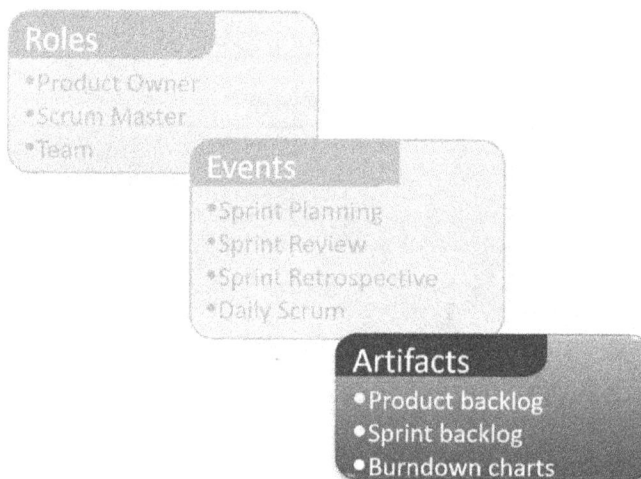

Roles
- Product Owner
- Scrum Master
- Team

Events
- Sprint Planning
- Sprint Review
- Sprint Retrospective
- Daily Scrum

Artifacts
- Product backlog
- Sprint backlog
- Burndown charts

CAPE PROJECT MANAGEMENT, INC.

Product Backlog

This is the product backlog

- ▸ The requirements
- ▸ A list of all desired work on the project
- ▸ Ideally expressed such that each item has value to the users or customers of the product
- ▸ Prioritized by the product owner
- ▸ Reprioritized at the start of each Sprint

55

Copyrighted materials. 2019

CAPE PROJECT
MANAGEMENT, INC.

Notes:

The Sprint backlog

- The Product Owner works with the development team to select Stories based upon a Sprint Goal
- Individuals sign up for work of their own choosing
- Any team member can add, delete or change tasks on the Sprint backlog
- Estimated work remaining is updated daily on backlog and burndown charts by the development team.

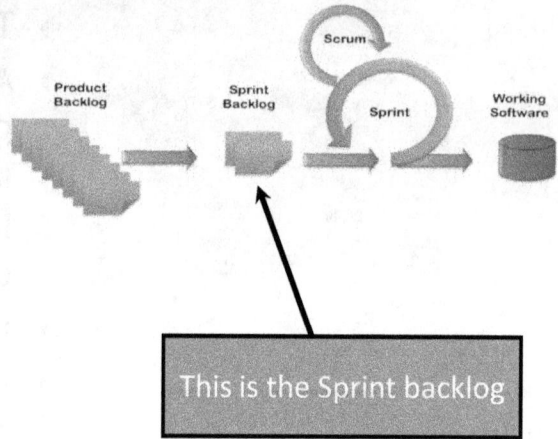

Product Backlog

Sprint Backlog

Scrum

Sprint

Working Software

This is the Sprint backlog

56

CAPE PROJECT
MANAGEMENT, INC.

Notes:

Burndown Charts

- ▸ Tracks work remaining
- ▸ At-a-glance information
  - ◦ Number of user stories committed
  - ◦ Duration of Sprint
  - ◦ Target velocity
  - ◦ Performance against plan
    - • Above the line– behind schedule
    - • Below the line – ahead of schedule
- ▸ Sprint burndown
- ▸ Release burndown
- ▸ Burndown bar chart

57

CAPE PROJECT
MANAGEMENT, INC.

Notes:

Velocity

- Velocity = Rate of Progress
- Number of story points to be completed in Sprint
- First Sprint is a guess
- Estimate improves over time
- Account for work done and disruptions on the project
- Based upon team synergy

58

Copyrighted materials. 2019

CAPE PROJECT
MANAGEMENT, INC.

Notes:

Sprint Burndown Chart
Work Remaining

Beginning of Iteration

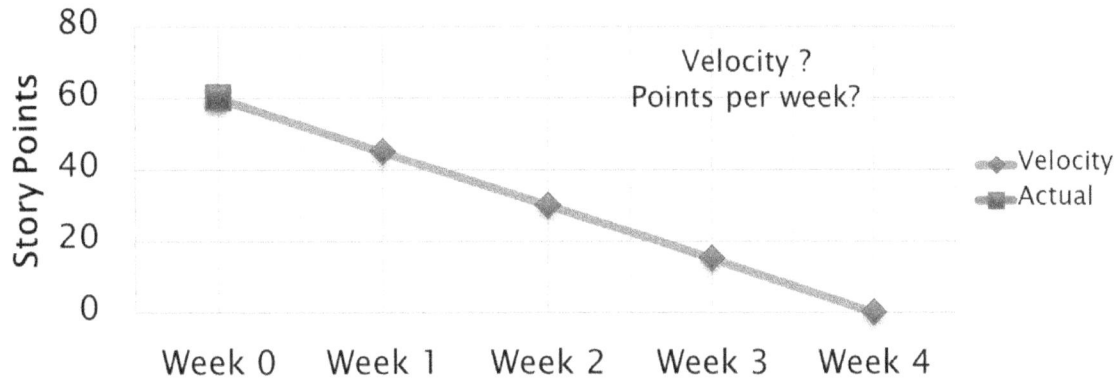

59

CAPE PROJECT
MANAGEMENT, INC

Notes:

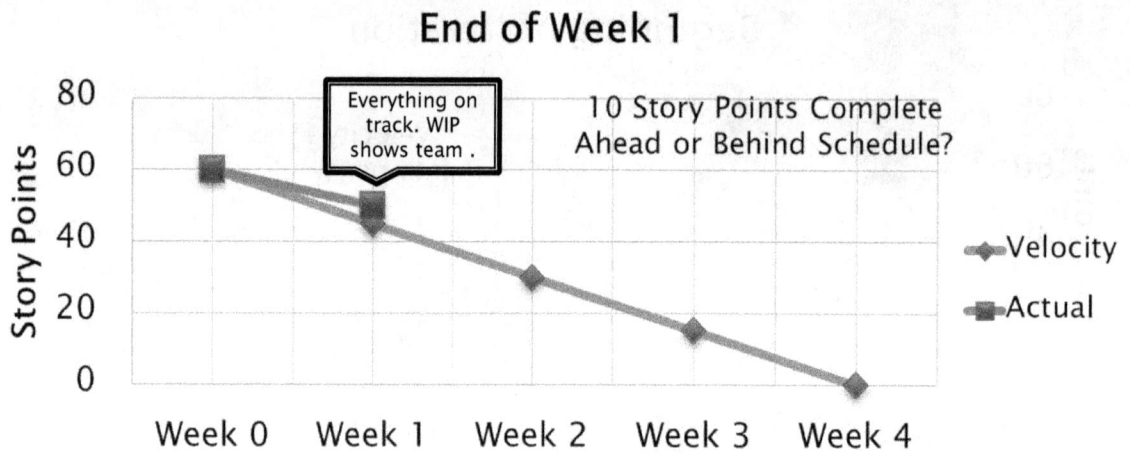

Notes:

| |
| |
| |
| |
| |
| |
| |
| |
| |

Sprint Burndown Chart

End of Week 2

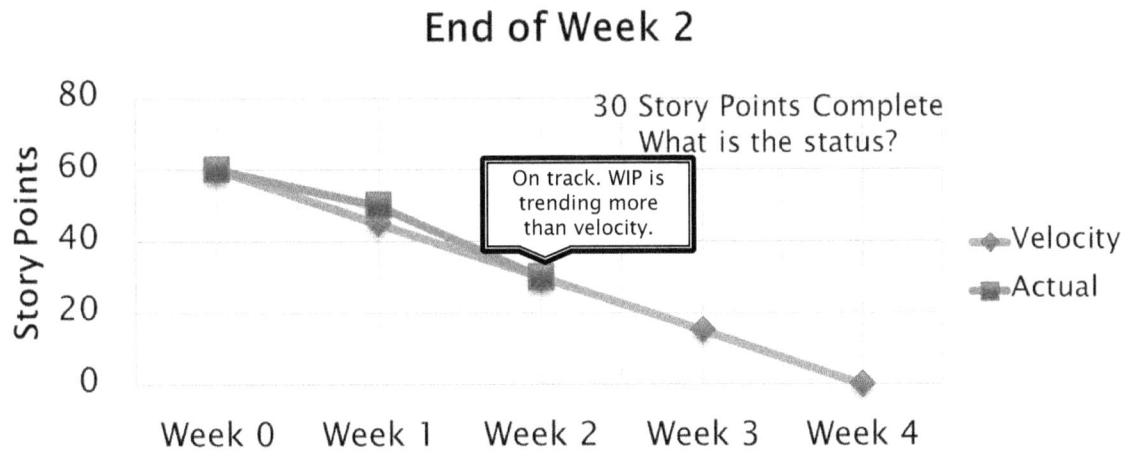

80

60

Story Points
40

20

0

Week 0   Week 1   Week 2   Week 3   Week 4

30 Story Points Complete
What is the status?

On track. WIP is trending more than velocity.

◆ Velocity
■ Actual

61

CAPE PROJECT
MANAGEMENT, INC.

Notes:

Sprint Burndown Chart

End of Week 3

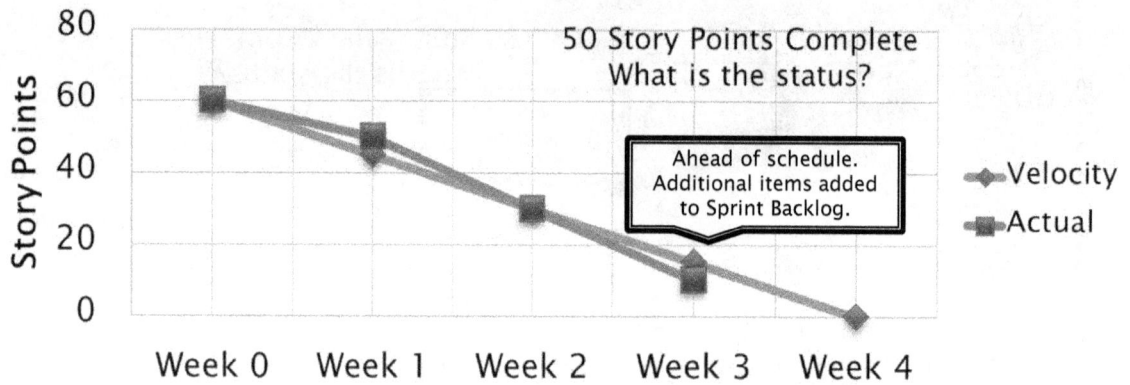

50 Story Points Complete
What is the status?

Ahead of schedule. Additional items added to Sprint Backlog.

Velocity
Actual

Week 0 Week 1 Week 2 Week 3 Week 4

Story Points

80
60
40
20
0

62

CAPE PROJECT MANAGEMENT, INC.

Notes:

Sprint Burndown Chart

End of Iteration/Sprint

63

Notes:

Forecasting Velocity for Release Planning

▸ **Rolling Average**
  ◦ Used for new teams
  ◦ 3-Sprint rolling average is most common
    • Sprint 1=30 points, Sprint 2= 26 points, Sprint 3= 40 points, Forecast Sprint 4 = 32 points

▸ **Yesterday's Weather**
  ◦ The last Sprint's velocity is the only "truth"
  ◦ Typically used in more mature teams

64

CAPE PROJECT
MANAGEMENT, INC.

Notes:

Using Velocity for Release Planning

Story Points

Velocity

=

Number of Sprints

CAPE PROJECT
MANAGEMENT, INC

Notes:

Release Burndown Chart

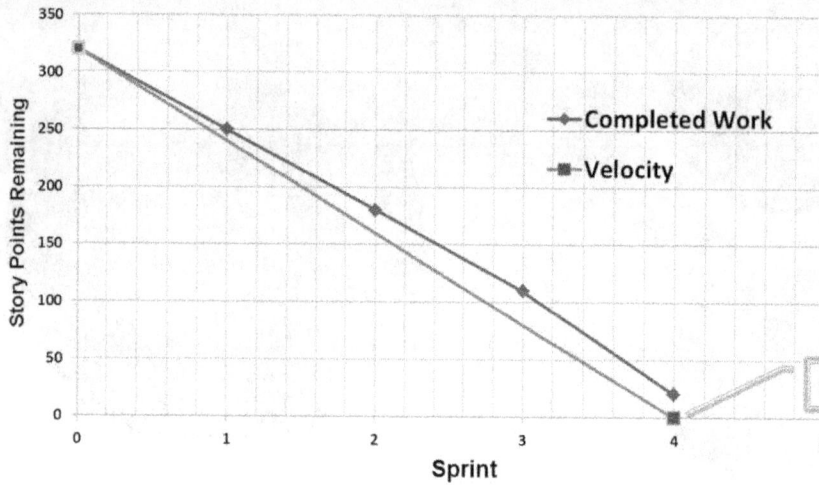

Backlog ?
Velocity ?
Sprints ?
Actual ?

Backlog = 320 pts
Velocity = 80
4 Sprints
Actual = 300 pts

Release

CAPE PROJECT
MANAGEMENT, INC.

Notes:

Scope, Date, Resource\Budget

A Case Study

CAPE PROJECT
MANAGEMENT, INC.

Notes:

Project Facts

- ▸ Experience with Scrum: 1st time ever!
- ▸ Backlog: 320 Points
- ▸ Sprint 1 Target: 20 Points
- ▸ Sprint duration: 2 weeks
- ▸ Project duration: 6 Months (12 Sprints)
- ▸ Budget: $585K
- ▸ Forecasted Budget: $673K

Do you see the problem?

Notes:

It's been 2 weeks:
How do we answer these Questions?

▸ Management wants to know when we will be done?

▸ Will we be able to complete all of the Scope?

▸ How would we recover from Sprint 1 if we underperform?

▸ How do we communicate lots of information in a simple manner that leaves no ambiguity?

▸ Let's take a look...

CAPE PROJECT
MANAGEMENT, INC.

Notes:

Sprint 1

- ▸ The team worked hard in Sprint 1, but came up short. They completed 10 of the 20 Points targeted.
- ▸ The developers pulled in a low Business Value story over a Higher Business Value story because it was "an easy one" and they were already in the code.
- ▸ While we didn't meet our goal, we still spent 2 weeks worth of effort (budget)!

CAPE PROJECT
MANAGEMENT, INC.

Notes:

3 Pie Charts

IT Points

10

320

- ◼ Total
- ◼ Done

Business Value Points

2400

15200

- ◼ Total
- ◼ Done

Budget

48,750

585,956

- ◼ Total
- ◼ Done

After 2 weeks we delivered 10 IT Points worth 2,400 Business Value Points for $48,750.00.

CAPE PROJECT
MANAGEMENT, INC.

Notes:

Sprint 1

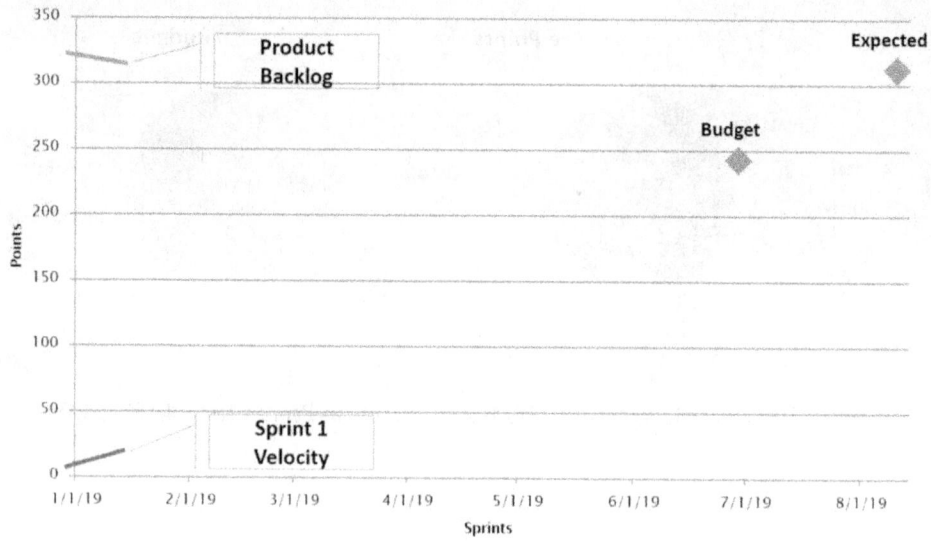

Notes:

Houston, we have a problem

- ▸ Sprint 1:
  - ◦ Team expected to deliver 20 points but only completed 10
- ▸ Sprint 2:
  - ◦ The Velocity increased, but the Team only completed 20 Points
- ▸ Today, the Business just informed us that Phase II requirements would be ready by May and they would take precedent over Phase I.
- ▸ We now have 290 (of the original 320) Points remaining to be completed in 10 Sprints.

What should we do?

CAPE PROJECT
MANAGEMENT, INC.

Notes:

Projections

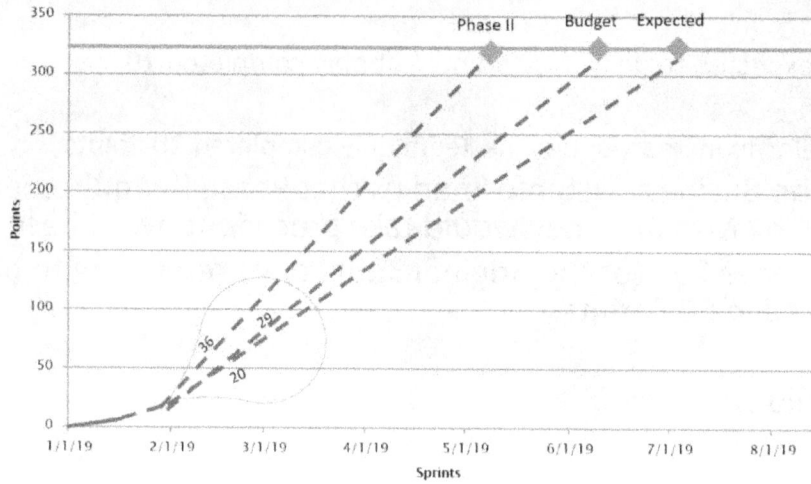

Copyrighted materials. 2019

CAPE PROJECT
MANAGEMENT, INC.

Notes:

Projections, a Closer look

- Empirical Evidence:
  - After Sprint 2, the Backlog was reduced by 30 Total Points
  - To complete the Backlog, the team would need 15 Sprints at their current Velocity when we only have 10 remaining.
    - 290/20 = 14.5 Sprints
- Budget:
  - To complete the Backlog on budget, the team would need to deliver 29 Points per Sprint for the remaining 10 Sprints.
    - 290/10 = 29 Points/Sprint
- Phase II:
  - To complete by May (one Month early) when Phase II requirements are delivered, the team would need to deliver 36 Points per Sprint.
    - 290/8 = 36 Points/Sprint

CAPE PROJECT
MANAGEMENT, INC.

Notes:

So, what happened?

Traditional Mindset:

- Hit the date! This means that the team has to hit 36 points/sprint to deliver by the Phase II milestone.

Agile Mindset:

- Scrum Master asks the team what they think they could do to meet the milestone.

The result?

- MVP – The Product Owner reviewed the 290 remaining points and defined a Minimum Viable Product that contained only 175 of the total points remaining required to go live.
- Now that the team was into Sprint 3, they started to understand their work habits better and increased performance.
- Better team work and by leveraging new automation tools, it allowed the team to increase their velocity from 20 to 29.

  CAPE PROJECT MANAGEMENT, INC.

Notes:

New Projections

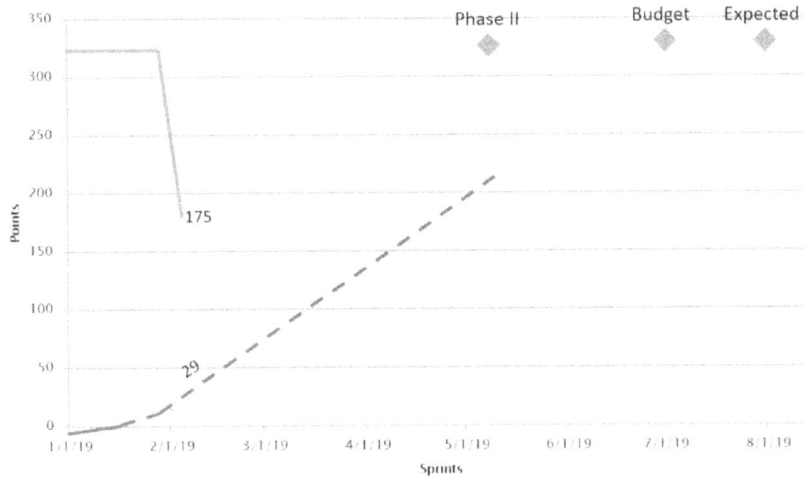

Notes:

Agile Dashboard

IT Points

54
266

■ Total
■ Done

Business Value Points

6400
15200

■ Total
■ Done

Budget

146,250
585,956

■ Total
■ Done

Phase II

After 3 weeks we delivered 54 IT Points worth 6,400 Business Value Points for $146,250

CAPE PROJECT
MANAGEMENT, INC.

Notes:

So, what can 2 weeks, 3 Pie charts and a graph can tell you?

▸ Empower the Team to figure things out and don't rely on Command and Control
▸ Make sure information is visible for all to see
▸ Make sure you use the information you have
▸ Don't lose sight of delivering Business Value!
▸ Deliver Incremental Value (MVP)

CAPE PROJECT
MANAGEMENT, INC.

Notes:

The Agile Product Lifecycle

Module 3

CAPE PROJECT
MANAGEMENT, INC.

Agile Planning Approach

Product or Project

What business objectives will the product fulfill?

Product Vision

Product Roadmap

Iteration/Sprint

What specifically will we build? (User Stories)

How will this iteration move us toward release objectives?

Sprint Goal

Sprint Backlog

(Circles, from outer to inner: Product Vision, Product Roadmap, Release Plan, Sprint Plan, Daily Plan)

Release

How can we release value incrementally?

What subset of business objectives will each release achieve?

What users will the release serve?

What general capabilities will the release offer?

Release plan

Product Backlog

Story (Backlog Item)

What user or stakeholder need will the story serve?

How will it specifically look and behave?

How will I determine if it's completed?

Story Details/Tasks

Acceptance Tests

Copyrighted materials. 2019

CAPE PROJECT MANAGEMENT, INC.

81

Notes:

Agile Plans

▸ Typically are top-down

▸ Are easy to change

▸ Limit dependencies

▸ Follow a rolling-wave approach

  ◦ Rolling Wave Planning is a technique that enables you to plan for a project as it unfolds. This technique requires you to plan iteratively. You plan until you have more visibility and then re-plan.

CAPE PROJECT
MANAGEMENT, INC.

Notes:

Agile Discovery

- **Desired Business Outcomes:** Document outcomes that are quantifiable and measurable.
- **High Level Architecture**: Outline a plan for the technical and business architecture/design of the solution.
- **High Level Delivery Plan or Agile Charter**: Segment the solution into the smallest minimum viable products (MVPs) that realize the desired outcomes and sets out the order in which they are to be delivered.
- **Initial Product Backlog**: Create an evolving prioritized list of all items of work which may be relevant to the solution.
- **Governance Approach**: Describe essential governance and organization aspects of the project and how the project will be managed.
- **High-Level Budget**: Create a project budget and release budget to support resource planning and scheduling.

CAPE PROJECT
MANAGEMENT, INC.

83

Notes:

Developing a Vision

▸ The product vision is key to the success of the project.

▸ The product vision should align with the company vision

▸ The vision should be revisited frequently

▸ All releases of the product should related back to the vision

Copyrighted materials. 2019

CAPE PROJECT
MANAGEMENT, INC.

Notes:

Product Roadmap

- A roadmap is a planned future, laid out in broad strokes
  - Planned or proposed product releases, listing high level functionality or release themes, laid out in rough timeframes
  - For a period usually extending for 2 or 3 significant feature releases into the future
- Shows progress towards strategy
- Lots of "wiggle room"
- Example:
  - Implement course listing functionality
  - Implement grading functionality
  - Implement discussion groups
  - Implement student profiles

85   Copyrighted materials. 2019

CAPE PROJECT
MANAGEMENT, INC.

Notes:

Product Roadmap Example

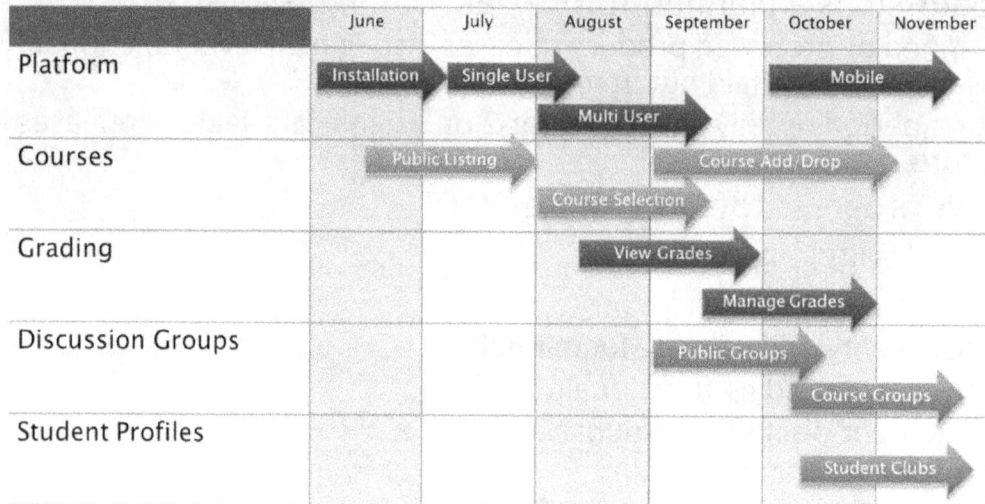

	June	July	August	September	October	November
Platform	Installation	Single User	Multi User		Mobile	
Courses		Public Listing	Course Selection	Course Add/Drop		
Grading			View Grades	Manage Grades		
Discussion Groups				Public Groups	Course Groups	
Student Profiles					Student Clubs	

86

Copyrighted materials. 2019

CAPE PROJECT
MANAGEMENT, INC.

Notes:

Release Planning

▸ Occurs once the vision and roadmap are complete

▸ Is created by the Scrum Team, stakeholders, project sponsors and customers when possible

▸ Identifies major feature releases for 3-6 months

▸ Each release should identify a minimum viable product (MVP)

87

CAPE PROJECT
MANAGEMENT, INC.

Notes:

Minimum Viable Product (MVP)

- The product with the highest return on investment versus risk.
- Just those core features that allow the product to be deployed, and no more.
- It allows you to test an idea by exposing an early version of your product to the target users and customers, to collect the relevant data, and to learn from it.

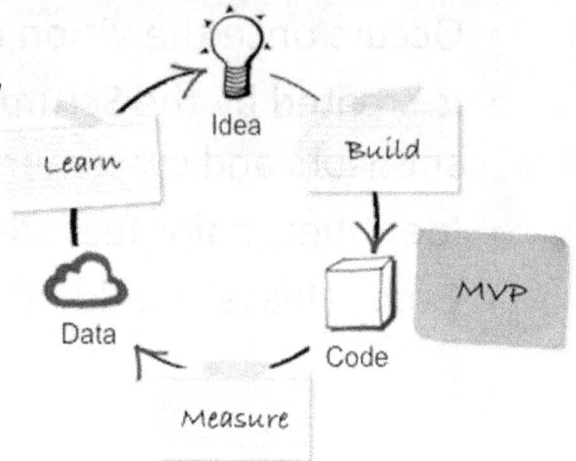

Source: http://www.romanpichler.com/blog/minimum-viable-product-and-minimal-marketable-product/

88

CAPE PROJECT
MANAGEMENT, INC.

Notes:

Requirements Management

The Product
Backlog

CAPE PROJECT
MANAGEMENT, INC.

89

Notes:

Product Backlog

▸ Ideally expressed such that each item has value to the users or customers of the product

▸ Aligns with the Product Roadmap and Release Plan

▸ As long as a product exists, its Product Backlog also exists.

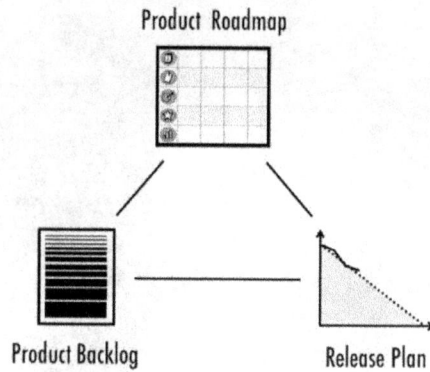

Product Roadmap

Product Backlog

Release Plan

http://www.romanpichler.com/blog/product-roadmap-vs-release-plan/

Copyrighted materials. 2019

CAPE PROJECT
MANAGEMENT, INC.

Notes:

Agile Requirements Hierarchy

| Vision |
| Roadmap |
| Release Plan |

| Theme | Theme |

| Epic | Epic | Epic | Epic |

| User Story | User Story |

| Task | Task |

CAPE PROJECT
MANAGEMENT, INC

Notes:

Questions

CAPE PROJECT
MANAGEMENT, INC.

Notes:

Agile
Controls

Module 4

CAPE PROJECT
MANAGEMENT, INC.

Agile Controls

▸ Change Management

▸ Quality Management

▸ Risk Management

CAPE PROJECT
MANAGEMENT, INC.

Notes:

Change Management in Agile

- ▸ Embrace it versus fight it
- ▸ Scrum—no change during a Sprint
- ▸ Product Owner owns change
- ▸ Change is just a reprioritization of backlog

CAPE PROJECT MANAGEMENT, INC.

95

Notes:

(blank lined area)

Agile Quality Assurance & Control

- ▶ **Product Owner/Customer in the Team**
  - ○ Software that meets the Product Owner's intent
  - ○ Product Owner is the single owner of quality and becomes part of the team and guides development.
- ▶ **Releasable Software in every Timebox, which:**
  - ○ Meets the Product Owner's value-driven approach
  - ○ Adheres to a Definition of Done
  - ○ Has been tested to the satisfaction of the team and relevant stakeholders
- ▶ **Iterative and Continuous Product Reviews**
  - ○ Results are documented and incorporated into backlog
  - ○ Defects are managed as backlog items
- ▶ **Definition of Done is the primary Quality Control tool**

96

CAPE PROJECT
MANAGEMENT, INC.

Notes:

Definition of "Done"

▸ The list of activities (coding comments, unit testing, integration testing, release notes, design documents, etc.) which supports the expected business value.

▸ The intention is to focus on value-added steps that allow the team to focus on what must be completed in order to build software while eliminating wasteful activities.

▸ The definition of "Done" is defined by the Product Owner and Development Team.

▸ Make "Done" more stringent over time with each release

97

CAPE PROJECT MANAGEMENT, INC.

Notes:

Risk Management in Agile

▸ Agile mitigates the most commons project risks:
  ◦ Intrinsic Schedule Flaw
  ◦ Specification Breakdown
  ◦ Scope Creep
  ◦ Personnel Loss
  ◦ Productivity Variance

▸ Risk management in Agile is "organic", since risk is addressed naturally as part of the project lifecycle.

▸ Qualitative risk analysis if preferred over traditional quantitative risk analysis to identify risk on Agile projects

98

Copyrighted materials. 2019

CAPE PROJECT
MANAGEMENT, INC.

Notes:

Do You Recognize These Risks?

Risk	Agile Mitigation
Misunderstanding of the requirements	Short delivery cycles with continuous feedback
Lack of management commitment and support	Stakeholder involvement throughout
Lack of adequate user involvement	Users actively engaged as a team member
Failure to gain user commitment	User creates the requirements
Failure to manage end user expectation	Prioritization by business value
Changes to requirements	Embrace change
Lack of an effective project management methodology	Repeatable lightweight methodologies

99   Copyrighted materials. 2019   CAPE PROJECT MANAGEMENT, INC.

Notes:

Questions

100

CAPE PROJECT
MANAGEMENT, INC.

Notes:

Implementing Agile

Module 5

CAPE PROJECT
MANAGEMENT, INC.

Top Reasons Agile Projects Fail

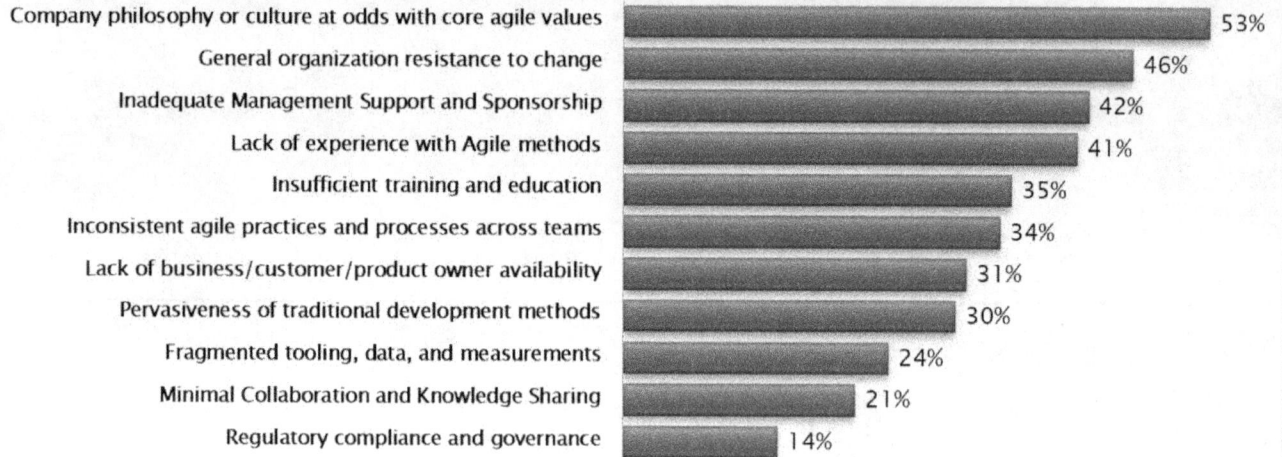

Reason	Percentage
Company philosophy or culture at odds with core agile values	53%
General organization resistance to change	46%
Inadequate Management Support and Sponsorship	42%
Lack of experience with Agile methods	41%
Insufficient training and education	35%
Inconsistent agile practices and processes across teams	34%
Lack of business/customer/product owner availability	31%
Pervasiveness of traditional development methods	30%
Fragmented tooling, data, and measurements	24%
Minimal Collaboration and Knowledge Sharing	21%
Regulatory compliance and governance	14%

Source: Version One 12th Annual Agile Survey, 2018

CAPE PROJECT
MANAGEMENT, INC.

102

Notes:

Notes:

Implementation Basics

- Acknowledge that "the way we were working wasn't working"
- Treat the implementation of Agile as a fundamental change to the organization
- Start with the basics then continuously improve

CAPE PROJECT
MANAGEMENT, INC.

Notes:

Activity

Force Field Analysis

105

CAPE PROJECT
MANAGEMENT, INC.

Notes:

Force Field Analysis

Directions

1. Use the worksheet on the next page.

2. On the center box, write taking change you are anticipating.

3. List all the forces FOR CHANGE in one column, and all the forces AGAINST CHANGE in another column.

4. Rate the strength of these forces and assign a numerical weight, 1 being the weakest, 5 being the strongest.

5. When you add the "strength points" of the forces, you'll see the viability of the proposed change.

The tool can be used to help ensure the success of the proposed change by identifying the strength of the forces against the change.

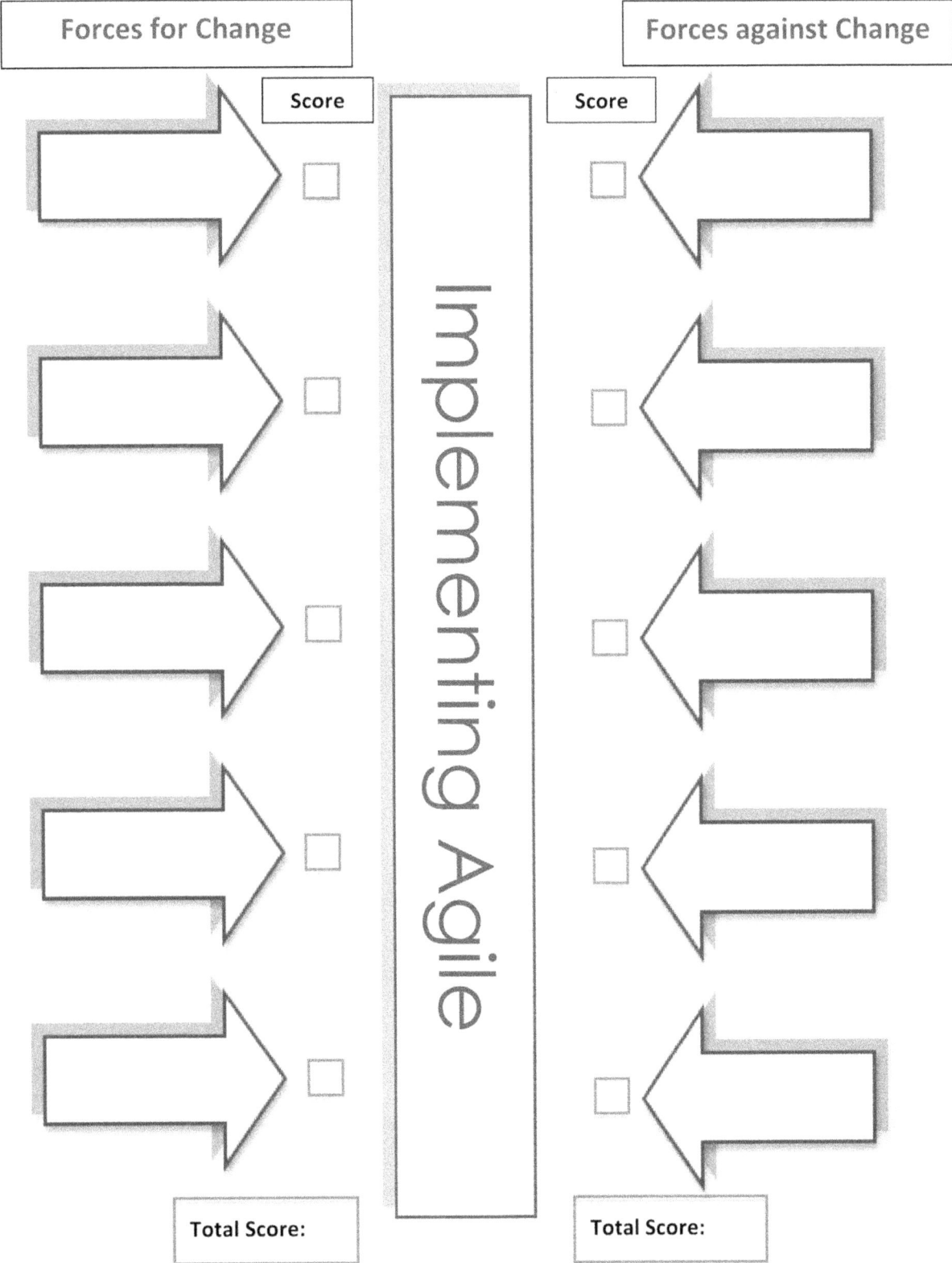

Forces for Change		Forces against Change
Score	Implementing Agile	Score

Total Score:

Total Score:

Questions

CAPE PROJECT
MANAGEMENT, INC.

Notes:

Thank-You!

dan@CapeProjectManagement.com

Twitter: @scrumdan

CAPE PROJECT
MANAGEMENT, INC.